AF270460

WILD STEM

STEM on the FARM

Megan Borgert-Spaniol

Checkerboard
Library

An Imprint of Abdo Publishing
abdobooks.com

abdobooks.com

Published by Abdo Publishing, a division of ABDO, PO Box 398166, Minneapolis, Minnesota 55439. Copyright © 2024 by Abdo Consulting Group, Inc. International copyrights reserved in all countries. No part of this book may be reproduced in any form without written permission from the publisher. Checkerboard Library™ is a trademark and logo of Abdo Publishing.

Printed in the United States of America, North Mankato, Minnesota
102023
012024

Design: Denise Hamernik, Mighty Media, Inc.
Production: Mighty Media, Inc.
Editor: Anna Anderhagen
Cover Photographs: klebercordeiro/Getty Images
Interior Photographs: Akarawut/Shutterstock Images, p. 5; Dmitry Chulov/Shutterstock Images, p. 11; Filip Ilic Novi Sad/Shutterstock Images, p. 19; macondo/Shutterstock Images, p. 13; Manop Boonpeng/Shutterstock Images, p. 29; Mark Brandon/Shutterstock Images, p. 9; Mighty Media, Inc., pp. 16–17, 22–23; Oleggg/Shutterstock Images, p. 15; Olena Znak/Shutterstock Images, p. 25; Phovoir/Shutterstock Images, p. 27; Srijaroen/Shutterstock Images, p. 21; Tanya Kalian/Shutterstock Images, p. 7
Design Elements: Francois Poirier/Adobe Stock; Lorthois Yuliya/Shutterstock Images; supanut/Adobe Stock

Library of Congress Control Number: 2023939337

Publisher's Cataloging-in-Publication Data
Names: Borgert-Spaniol, Megan, author.
Title: STEM on the farm / by Megan Borgert-Spaniol
Description: Minneapolis, Minnesota : Abdo Publishing, 2024 | Series: Wild STEM | Includes online resources and index.
Identifiers: ISBN 9781098291990 (lib. bdg.) | ISBN 9781098278892 (ebook)
Subjects: LCSH: Science--Study and teaching--Juvenile literature. | Technology--Study and teaching--Juvenile literature. | Engineering--Study and teaching--Juvenile literature. | Mathematics--Study and teaching--Juvenile literature. | Farms--Juvenile literature. | Rural life--Juvenile literature.
Classification: DDC 577.0--dc23

CONTENTS

WILD FARM

Do you love watching rows of corn sprout up through the soil? Do you dream of raising cows, chickens, or pigs? If so, you might enjoy a career that takes you to the farm!

Many people work on farms. Livestock and poultry farmers raise animals for their meat, milk, and eggs. Fish farmers produce fish and seafood. Flower farmers harvest beautiful blooms, while beekeepers manage hives for their honey. Crop farmers grow plants for food, clothing, and more.

What kind of farmer would you like to be? There's only one way to find out. Put on your hat and grab some work gloves. It's time to explore science, technology, engineering, and math (STEM) on the farm!

STEM
Science
Technology
Engineering
Math

By 2050, farmers will have to produce about
60 percent more food than they do currently to
feed the world's population!

LIVESTOCK & POULTRY FARMERS

When was the last time you ate a hamburger, fried an egg, or wore a wool sweater? All these activities are possible because of livestock and poultry farmers. Livestock farmers raise animals such as cattle, pigs, and sheep. These animals provide products including milk, meat, and wool. Poultry farmers raise chickens, turkeys, and other birds for meat and eggs.

Livestock and poultry farmers may manage the care of hundreds of animals! Technology helps them attend to every animal. Farmers keep track of individual animals with electronic identification tags. These electronic tags hold information about each animal's eating habits, movements, and health records.

Some livestock farmers track their animals using technology that recognizes each animal's face!

Farmers also use technology to perform daily tasks. For example, many farmers use robotic milking computer systems. The computer keeps track of when the cow was last milked and when it's time to milk again. The robot cleans the cow's udder and attaches the milk pumps. The robot then milks the cow and releases it from the stall.

In addition to milking cows, robots collect eggs. They also wash pig stalls, **sanitize** chicken houses, and even clean up cow poop! Technologies like these help keep farm facilities clean and animals healthy.

Robotic milking saves farmers time and money. Dairy farmers must milk each cow two or three times a day.

FISH FARMERS

Most farmers raise animals on land, but fish farmers raise animals in water. The farming of fish, shrimp, and other seafood is called aquaculture. Aquaculture provides food for humans while protecting wild fish populations.

Some fish farmers raise fish in offshore pens in lakes or oceans. Other fish farmers raise fish in ponds or indoor tanks. Offshore pens can spread chemicals and parasites to the wild fish populations around them. Ponds and indoor tanks avoid these problems because they separate farmed fish from wild **habitats**.

New technologies aim to make fish farming more **sustainable**. Advanced **sensors monitor** water pollutants and parasites in offshore pens. Solar panels power wastewater treatment in ponds and tanks. Tools like these help fish farmers feed the world!

Many fish farmers grow oysters and other shellfish on beaches (*pictured*) or suspended in water.

BEEKEEPERS

Honey is the sweetest animal product that humans consume. Honeybees produce honey. Farmers who raise honeybees are called beekeepers.

Beekeepers protect honeybee colonies in structures called beehives. Beekeepers create beehives out of stacked wooden boxes. Honeybees lay eggs in the lower compartments and store honey in the upper compartments.

When beekeepers place a colony in a hive, the bees get to work building honeycomb. Honeycomb is a structure of cells made of wax from the honeybees' bodies. The queen bee lays eggs in this honeycomb.

Worker bees collect nectar from flowers near the hive. They store the nectar in honeycomb cells in the upper part of the hive. There, the water from

Beekeepers wear hats with veils to protect their faces from bee stings. Some wear suits for full-body protection.

the nectar evaporates, leaving behind honey. The worker bees cap the honey-filled honeycomb with wax.

Honeybees store honey to feed the colony. But they produce more honey than the colony needs. A bee colony can produce 50 to 100 pounds (23 to 45 kg) of excess honey in a year! One of the most important duties of a beekeeper is to harvest this extra honey.

To harvest honey, beekeepers pull frames from the upper part of a hive. Then, they place the hive frames into an **extractor**. This machine spins fast to force honey out of the honeycomb cells. Then, the honey is jarred and ready for sale. The extra wax from the honeycomb cells, called beeswax, is used to make lip balms, candles, and other products!

Beekeepers use tools called smokers when opening hives. The smell of smoke prevents bees from stinging to defend the hive.

GET WILD!

BEE SHELTER

Use household materials to create a safe shelter for solitary bees!

WHAT YOU NEED:

- ✔ paints & paintbrushes
- ✔ empty aluminum can
- ✔ hammer & nail
- ✔ twine
- ✔ ruler
- ✔ scissors
- ✔ cardboard toilet paper tubes
- ✔ tape

1. Use paints to decorate the can. Ask an adult to use a hammer and nail to make two holes in the side of the car, one near the top and one near the bottom. String about 2 feet (0.6 m) of twine through the holes. Knot the ends.

2. Open a toilet paper tube by cutting in a straight line along its entire length. Cut this piece of cardboard in half along the entire length again.

3. Roll up the two cardboard pieces into new tubes with an opening about ¼ inch (0.6 cm) wide. Use tape to secure the rolled-up cardboard. Make sure each tube you create is shorter than the length of the can. Cut off any extra length. Then place the cardboard tubes into the can.

4. Repeat steps 2 and 3 to make more tubes until the can is full.

5. Hang the bee shelter securely in a sunny spot at least 3 feet (0.9 m) above the ground. The opening of the can should face south or east so it gets the morning sun.

CROP FARMERS

Do you eat pasta, bread, or rice? Do you wear cotton T-shirts and jeans? Eating and wearing these things is possible because of crop farmers!

Food crops are grown to be eaten by humans. Grains are the most common food crop around the world. Popular grains include wheat, rice, and corn. Feed crops are grown to feed animals. These crops include oats, alfalfa, and grasses.

Fiber crops, such as cotton, are used to make clothing and paper products. Oil crops create oils we use for cooking. They also produce oils for paints, soaps, and fuel. Oil crops include soybeans, sunflowers, and olives.

Most crop farmers use similar equipment. They use plows to lift and turn the soil to prepare fields for planting. Then, farmers use planter machines to perform

A carrot harvester pulls carrots from the ground.
The average American eats more than eight pounds (3.6 kg)
of fresh carrots in a year!

several tasks at once. Planters cut into the ground, drop seeds, and cover the seeds as they move across the field.

As crops grow, farmers **irrigate** and fertilize the plants. If needed, they spray chemicals to get rid of harmful insects or weeds. When the plants are grown, crop farmers use machines called harvesters to cut or uproot them.

More advanced tools allow farmers to grow more crops with less human labor. Robots pick ripe strawberries from fields. **Autonomous** tractors operate without human drivers. Solar-powered drones fly over fields to fertilize plants and spray weeds. Tools like these will help farmers feed the world's growing population!

A farmer flies an agricultural drone to spray fertilizer on his rice field.

GET WILD!

STARTING SEEDS

Do you want to harvest your own food crops?

Use a cardboard egg carton to start your plants from seed. Then **transplant** your seedlings and watch your crops grow!

WHAT YOU NEED:

- ✔ cardboard egg carton
- ✔ scissors
- ✔ sharp pencil
- ✔ plate
- ✔ potting soil
- ✔ seeds (try peas, beans, or radishes)
- ✔ spray bottle
- ✔ water

1. Cut the cover off the egg carton and cut the egg carton into individual cups. Use the pencil to poke three holes into the bottom of each egg cup.

2. Place the egg cups onto the plate. Fill each egg cup with potting soil.

3. Use your finger to make a small well in the soil of each cup. Place one to three seeds in each well and cover them with soil.

4. Place the soil cups in a warm, sunny spot indoors. Water each cup regularly with the spray bottle to keep the soil damp.

5. Watch for seeds to sprout and grow into seedlings. When a seedling grows taller, plant it with the egg cup into a larger pot or in the ground. The cardboard egg cup will break down in the soil!

FLOWER FARMERS

Your grocery store contains farm products. You'll find them in the dairy aisle, in the produce section, or at the meat counter. You might also see a display of colorful flowers. Farmers grow flowers too!

Some flower farms are the size of a small park. Others are more than 60 acres (24 ha). That's the size of about 45 football fields! The largest flower farms grow millions of flowers each year.

Some flower farmers plant seeds or bulbs in the ground. Other farmers plant seeds in raised flower beds. Large flower farms grow their plants in greenhouses. These structures maintain the temperature and **humidity** that plants need to thrive year-round.

The Netherlands is the world's top flower producer. It is known for its tulips. Other leading flower producers include Colombia, Ecuador, and Kenya.

Many flower farmers use drip **irrigation** to water their plants. This system uses hoses that run along the ground. These hoses deliver water to the plants' roots slowly but steadily.

When flowers are ready for harvest, most farmers hand-cut the stems with sharp shears. Cut flowers must be kept cool so they don't wilt. Farmers move their cut flowers to a cooler before packaging them. Large flower producers ship their flowers over land and sea in refrigerated containers.

Large farms sell their products to **wholesale** flower companies around the world. Small farms sell their products at local grocery stores and florists. Wherever they sell, flower farmers spread color with their beautiful products!

Some flowers are farmed for food, not just for beauty.
Sunflowers are mostly grown for their oil and seeds.

A CAREER ON THE FARM

Do you have a passion for STEM on the farm? There are many career paths to choose from. Each path leads to a wide range of opportunities.

Some poultry farmers manage thousands of chickens on **industrial farms**. Other farmers raise a dozen chickens that freely roam a small plot of land. Some crop farmers use manufactured chemicals to fertilize and kill pests. Other farmers run organic farms that don't use these chemicals.

Some people work on a farm directly. Other people work as a scientist or engineer creating **innovations** that farmers use. Farmers and scientists of all kinds work hard to produce quality products for **consumers** worldwide!

Maybe one day, the farm will be your office!
What will you discover?

GLOSSARY

autonomous—relating to a device capable of operating without direct human control.

consume—to buy and use goods. A consumer is a person who buys and uses goods.

extractor—a machine that withdraws something by a physical process.

habitat—a place where a living thing is naturally found.

humidity—the amount of moisture in the air.

industrial farm—a farm that grows animals or plants in large quantities.

innovation—a new idea, method, or device.

irrigate—to water plants by artificial means to foster plant growth. This process is called irrigation.

monitor—to watch, keep track of, or oversee.

sanitize—to keep something free from disease by cleaning it.

sensor—an instrument that can detect, measure, and transmit information to a controlling device.

sustainable—relating to a method of using a resource so that the resource is not used up or damaged.

transplant—to move something from one place to another.

wholesale—having to do with selling something for resale.

ONLINE RESOURCES

To learn more about STEM on the farm, visit **abdobooklinks.com**. These links are routinely monitored and updated to provide the most current information available.

INDEX